CELEBRATING PHYSICS
WITH POETRY

CELEBRATING PHYSICS WITH POETRY

WALTER THE EDUCATOR

Silent King Books

dedicated to all the Physics lovers in the universe

Contents

Why I Created this Book?

Creating a poetry book that celebrates the subject of Physics can be a unique and enriching experience. Physics, often seen as a complex and technical subject, can be brought to life through the beauty of poetry. By combining the elegance of language and the depth of scientific concepts, this book can bridge the gap between art and science, appealing to a wider audience.

A poetry book on Physics can serve several purposes. Firstly, it can make the subject more accessible and relatable to readers who may not have a scientific background. Through creative and imaginative language, complex theories and ideas can be simplified and presented in a way that sparks curiosity and wonder.

Secondly, a poetry book can capture the essence and spirit of scientific exploration. Physics is all about understanding the fundamental principles that govern the universe, and poetry can capture the awe-inspiring nature of the subject. It can delve into topics like the mysteries of the cosmos, the intricacies of quantum mechanics, or the laws that shape our daily lives.

Additionally, a poetry book on Physics can help to highlight the creative and artistic aspects of scientific inquiry.

Science and art are not mutually exclusive; they both require imagination, observation, and creative thinking. By intertwining scientific concepts with poetic expression, the book can showcase the beauty and creativity inherent in scientific exploration.

One

unlocking the universe

Where brilliance resides,
There's a subject that unravels the universe's tides,
Physics, the mistress of forces unseen,
Unveiling the secrets of what lies in between.

From the tiniest atoms to the vast cosmic expanse,
Physics explores the laws of this cosmic dance,
Newton's laws, a foundation so strong,
Guiding us through motions, righting every wrong.

Einstein, the genius, with his theory of relativity,
Transforming our perception of time and activity,
Black holes and wormholes, bending space,
Physics takes us to extraordinary realms, in a chase.

Quantum mechanics, mysterious and strange,
Particles entangled in a quantum change,
In wave-particle duality, they dance and sway,
Uncertainty reigning in this quantum display.

From the laws of thermodynamics, we learn,
Energy's flow, in every twist and turn,
Entropy, a measure of disorder and change,
Physics unveils the world's constant exchange.

In laboratories, experiments take flight,
Particles collide, unveiling the light,
Accelerators and colliders, pushing the bounds,
Physics explores the smallest particles, profound.

From the grandeur of the cosmos to the subatomic scale,
Physics unravels the mysteries, without fail,
A symphony of equations, elegant and precise,
Guiding us through nature's every device.

So let us celebrate this noble pursuit,
The wonders of Physics, forever astute,
For in its realm, we find answers profound,
Unlocking the universe, in leaps and bounds.

Two

language of the cosmos

Wonders abide,
There's a subject that fills me with unbridled pride.
Physics, the mistress of matter and space,
Unveiling the secrets of our cosmic embrace.

With equations and laws, it unravels the unknown,
From the tiniest particles to the grandest cyclone.
From Newton to Einstein, the giants of thought,
Their genius and brilliance, the world sought.

In the realm of Physics, we venture through time,
Peering into the cosmos with a curious mind.
Black holes and quasars, pulsars that spin,
Revealing the mysteries that lie deep within.

We study the forces, the energy in flow,
The dance of atoms, in a world we can't see or know.
Magnetism, electricity, and waves in the air,
Creating a symphony, beyond compare.

From relativity's bend to quantum's strange ways,
Physics takes us on a journey that forever stays.
Unifying the laws of the universe, it strives,
To understand how everything around us survives.

From the birth of the stars to the depths of the sea,
Physics guides us to a deeper reality.
With every discovery, our knowledge expands,
Unlocking the secrets with its guiding hands.

So let us celebrate the wonders it brings,
The intricate beauty that it constantly sings.
Physics, the language of the cosmos above,
Forever inspiring, with its boundless love.

Three

a symphony of particles

A symphony of particles, from the grand to the small,
Physics, the language that explains it all.
From the twinkling stars that light up the night,
To the forces that keep us grounded with might.
Gravity pulls, while light travels fast,
Physics unravels the secrets, unsurpassed.

In the realm of atoms, where mysteries lie,
Quantum mechanics makes the particles fly.
Electrons whirling, in orbits unseen,
Uncertainty reigns, in this quantum machine.

Energy transforms, from one form to another,
Conservation of energy, a law to uncover.
Thermodynamics reveals how heat flows,
From hot to cold, energy constantly grows.

In the vast cosmos, where galaxies roam,
Astrophysics unveils the secrets of the unknown.

Black holes devour, while stars burst in light,
The universe expands, a cosmic delight.
From the laws of motion to the speed of sound,
Physics teaches us to explore, to astound.
From the simple pendulum to the complex machine,
Physics unlocks the wonders, yet unseen.
So let us celebrate, with wonder and awe,
The beauty of Physics, its limitless draw.
For in this realm, where knowledge unfolds,
Physics guides us, as our story unfolds.

Four

scientific journey

Laws and energy,
Where atoms dance in perfect symmetry,
There lies a realm of truth untold,
A world of wonder, where secrets unfold.

Physics, the muse of curious minds,
Unraveling the mysteries that bind,
From Newton's apple to Einstein's relativity,
A journey through the depths of creativity.

In the quantum realm, particles entangle,
Where probabilities and uncertainties wrangle,
From Schrödinger's cat to Heisenberg's uncertainty,
A tapestry of quantum complexity.

The laws of motion, they govern all,
From the humblest leaf to the mightiest fall,
For every action, there's a reaction,
A symphony of forces in perfect interaction.

Through lenses and mirrors, light bends and refracts,
In waves and particles, it beautifully acts,
From the rainbow's hues to the distant stars,
Physics unveils the cosmos, no matter how far.

From black holes to cosmic strings,
The universe's symphony, it sings,
With gravity's pull and time's dilation,
Physics paints a portrait of cosmic creation.

So let us marvel at the wonders of Physics,
From the tiniest quark to the grandest cosmics,
With every equation and every discovery,
We celebrate the beauty of this scientific journey.

Five

grand expanse

Atoms, and the electrons dance,
Physics, the language of cosmic romance.
From particles small to the vast expanse,
A symphony of forces, a cosmic advance.

Newton's laws guide motion with grace,
Gravity's pull, a celestial embrace.
Einstein's relativity, time's bending race,
Quantum mysteries, a quantum chase.

From the stars that shimmer in the night
To the planets spinning in their flight,
Physics unveils the secrets of light,
Unraveling nature's tapestry, shining bright.

In laboratories, experiments unfold,
Particles collide, stories yet untold.
Mysteries unravel, as theories take hold,
Physics, a journey, where wonders unfold.

From the tiniest quark to the vast black hole,
Physics unravels the universe's role.
Through equations and theories, we seek to know,
The secrets of existence, how it all bestows.

So let us celebrate the world of Physics,
A subject that illuminates and mystifies us.
With every discovery, our knowledge uplifts,
Physics, the canvas where reality thrusts.

In the realm of atoms, electrons dance,
Physics, the language of cosmic romance.
From particles small to the vast expanse,
We marvel at the wonders of this grand expanse.

Six

the joy it sings

Forces intertwine,
Where matter dances, laws align,
Lies a realm of endless wonder,
Where Physics reigns, a spell we're under.

From the atom's core to galaxies vast,
Physics unveils the secrets amassed,
It unravels the fabric of space and time,
And beckons us to ponder the sublime.

In equations and theories, it finds its voice,
Unifying the particles, giving them choice,
From Newton's laws to Einstein's might,
Physics illuminates nature's light.

With every experiment, a quest unfolds,
From laboratories to stories untold,
Exploring quantum realms and relativity,
Physics paints a picture of our reality.

It measures the speed of light's embrace,
Calculates the path of planets in space,
It unriddles the mysteries of the universe,
With formulas that make one's mind traverse.

From electricity's hum to the magnet's pull,
Physics reveals the forces that rule,
It gives birth to technology's advance,
And fuels our dreams with every chance.

So let us celebrate the wonders it brings,
The knowledge it imparts, the joy it sings,
For in the realm where Physics resides,
A tapestry of knowledge forever abides.

Seven

Inviting us to explore

Physics, the language of the universe profound,
Unveiling mysteries, with knowledge unbound.
From Newton's apple to Einstein's relativity,
Physics reveals nature's grandiosity.
In the realm of motion, forces hold sway,
Gravity pulls, while friction holds at bay.
Energy flows, in waves it does soar,
From electromagnetic to sound, we explore.
Quantum particles, elusive and strange,
In their dual nature, a paradox they arrange.
In laboratories, with intricate machines,
Scientists delve into quantum dreams.
Particles collide, revealing secrets untold,
The quarks and leptons, mysteries unfold.
From the vastness of space to the depths of the Earth,
Physics unveils the universe's birth.

Black holes devour, while stars brightly ignite,
Cosmic symphonies, dancing in the night.
   From the smallest particles to the galaxies afar,
Physics connects all, like a cosmic memoir.
From the laws of thermodynamics to optics of light,
Physics reveals the world, in all its might.
   So let us celebrate this wondrous science,
That unlocks the secrets with sheer reliance.
Physics, the muse of the curious mind,
Inviting us to explore, and knowledge to find.

Eight

❦

awe and grace

    A subject reigns supreme,
A realm where mysteries unravel, it's a gleam.
Physics, the art of understanding the laws,
That govern the universe, with no pause.

    From the tiniest particles, infinitesimally small,
To the vastness of galaxies, standing tall.
Physics delves into the essence of it all,
Unraveling nature's secrets, both big and small.

    Oh, the wonders of motion, in its graceful dance,
As objects glide and twirl, in perfect balance.
From Newton's laws to Einstein's relativity,
Physics unveils the marvels of connectivity.

    Electricity, magnetism, forces that bind,
Invisible powers, yet their effects we find.
With Maxwell's equations, we comprehend,
The language of light, a symphony to transcend.

The quantum realm, where particles reside,
In strange superpositions, their fate implied.
Heisenberg's uncertainty, a perplexing enigma,
Yet in its puzzles, lies the beauty of sigma.

From thermodynamics to waves and sound,
Physics captures phenomena all around.
With equations and formulas, it quantifies,
The mysteries of nature, it demystifies.

In labs and classrooms, minds ignite,
Exploring the frontiers, reaching new heights.
Physics, a beacon of scientific delight,
Guiding us toward a future, shining bright.

So let us celebrate this noble field,
Where knowledge and wonder are revealed.
Physics, the language of the universe, we embrace,
A subject that leaves us in awe and grace.

Nine

connects our hearts

Atoms dance and energies collide,
A symphony of laws and forces reside.
A subject profound, mysterious, and grand,
Physics, the language of the universe, we understand.
From the tiniest particles to the vast cosmic space,
Physics unravels the secrets, with infinite grace.
Gravity's pull, a force we all know,
Binding celestial bodies, making planets glow.
Einstein's relativity, a theory so bold,
Unveiling the nature of time, yet untold.
With black holes and wormholes, a cosmic ballet,
Physics paints a picture, in its own unique way
Quantum mechanics, a realm of the small,
Where particles entangle, enthralling us all.
Superposition and uncertainty, a puzzling quest,
Physics pushes boundaries, putting theories to the test.

In labs and observatories, scientists explore,
Unraveling the mysteries, seeking to know more.
From Galileo to Hawking, minds so bright,
Physics transcends boundaries, reaching out to the light.

From the laws of motion to electromagnetic waves,
Physics weaves a tapestry, the universe it engraves.
From the birth of stars to the gentle breeze,
Physics whispers its secrets, upon the cosmic seas.

So let us celebrate this subject divine,
Physics, the language that makes the world align.
With equations and experiments, we seek to find,
A deeper understanding, the wonders of humankind.

In the realm of Physics, we stand in awe,
A symphony of knowledge, a quest to explore.
So let us embrace the beauty it imparts,
For Physics, the language that connects our hearts.

Ten

guide us in this quest

Mysteries unfold,
Lies a subject of wonders untold.
Physics, the language of the universe,
Where laws are written to disperse.

From atoms dancing in the cosmic space,
To galaxies twirling in a cosmic embrace.
Physics reveals the secrets so grand,
Unveiling the workings of this vast land.

Oh, the laws of motion, so elegant and true,
Newton's legacy, forever imbued.
For every action, a reaction bestowed,
In this dance of forces, a story unfolds.

Electricity flows, electrons in flight,
Magnetism's pull, a force of might.
Maxwell's equations, harmonies divine,
Lightning and circuits, their secrets align.

Quantum mechanics, a realm so strange,
Particles entangled, in patterns they arrange.
Wave-particle duality, a cosmic paradox,
Where uncertainty reigns, in quantum blocks.

Einstein's theories, relativity's grace,
Bending spacetime, in its cosmic embrace.
Black holes devour, stars collapse,
A universe warped, in gravity's laps.

From subatomic realms to celestial spheres,
Physics unravels the mysteries that appear.
A symphony of equations, a cosmic score,
Revealing the universe's deepest core.

So let us celebrate this science profound,
Physics, the language where wonders abound.
Let curiosity guide us in this quest,
To understand the universe, its secrets blessed.

Eleven

space and time

From quarks to galaxies, it unveils the truth,
Unraveling mysteries, like an endless sleuth.
Einstein's theory, a revelation profound,
Relativity's embrace, where wonders abound.
Speed of light, time's elusive chase,
Bending space and time, in a cosmic embrace.
Newton's laws, the foundation of motion,
Gravity's pull, a universal devotion.
Inertia's might, objects at rest or in flight,
Velocity, acceleration, a symphony of might.
Quantum mechanics, a world unseen,
Particles entangled, caught in-between.
Wave-particle duality, a paradox untold,
Uncertainty's embrace, nature's secret hold.
From thermodynamics to electricity's dance,
Magnetism's allure, a magnetic romance.

Ohm's law, circuits ablaze with might,
Electromagnetism, a symphony of light.

Astrophysics, the study of celestial spheres,
Black holes devouring, time's flow unclear.
Stellar explosions, supernovae ignite,
Cosmic fireworks, illuminating the night.

From the tiniest particles to galaxies grand,
Physics, the language that helps us understand.
The laws of nature, a symphony divine,
Physics, the enchantress of space and time.

Twelve

glimpse of the divine

From the smallest particles to the vast expanse,
A symphony of forces, in every cosmic dance.
Oh, Physics! The language of the universe,
Where equations whisper and theories immerse.
From the laws of motion to the speed of light,
You guide us through the darkness, shining bright.

In the realm of matter, you unveil the truth,
Unraveling secrets, from the days of youth.
From the laws of thermodynamics to quantum fields,
You show us the wonders that nature conceals.

Through the lens of optics, we witness the light,
Bending and reflecting, revealing its might.
From the colors of rainbows to celestial sight,
You paint the canvas of our scientific flight.

Oh, Physics! You stretch our minds to explore,
Challenging boundaries, forever seeking more.

From relativity's grace to black holes' might,
You take us on journeys, both day and night.

With every discovery, a new door swings wide,
Inviting us to explore, with passion as our guide.
From the mysteries of dark matter to the birth of stars,
Physics, you inspire us to reach for afar.

Oh, Physics! You're the conductor of the symphony,
Guiding us through the complexities of infinity.
With every equation, a glimpse of the divine,
A testament to the wonders that science can define.

So let us celebrate this subject of awe,
With wonder and curiosity, forever in awe.
For in the realm of Physics, we find our place,
A journey of discovery, a cosmic embrace.

Thirteen

nature's secrets

Physics is a vast domain,
A subject reigns, both awe-inspiring and arcane,
Physics, the mistress of nature's secrets,
Unraveling the mysteries that the universe begets.

With quivering atoms and dancing waves,
Invisible forces that the cosmos engraves,
From quantum realms to celestial spheres,
Physics unveils the wonders that ignite our fears.

From Newton's apple to Einstein's relativity,
The laws of physics shape our reality,
Gravity pulls, while light travels fast,
Atoms collide, creating energies unsurpassed.

In the depths of space and time's embrace,
Black holes devour, leaving no trace,
Particles collide, creating brilliant flashes,
Revealing the universe's cosmic clashes.

From the tiniest quarks to galaxies afar,
Physics explores both near and far,
It seeks to understand the fundamental laws,
That govern the universe, without a pause.

Matter and energy, intertwined in a dance,
Einstein's equations provide a glance,
At the fabric of space, where time does bend,
Physics delves into realms that transcend.

So let us celebrate this noble pursuit,
Where curiosity and knowledge take root,
Physics, the gateway to nature's grand design,
A subject that continues to astound, sublime.

Fourteen

forevermore

A symphony of forces, unseen and profound,
In the realm of matter, wonders are found.
Oh, Physics! Thy laws are a cosmic guide,
From the tiniest quark to the stars so wide.
With equations and theories, you unravel the unknown,
Unveiling the mysteries, like a stellar throne.
The world of particles, a tapestry of charm,
Fermions and bosons, in a quantum swarm.
From protons to neutrinos, they dance and they spin,
In a cosmic ballet, where new worlds begin.
Electricity crackles, a thunderous might,
Magnetism pulls, with a force so bright.
Electromagnetic waves, in a cosmic ballet,
Revealing the wonders, of night and day.
Gravity, the weaver of cosmic threads,
Bends space and time, with invisible threads.

Planets orbit, in celestial embrace,
Guided by gravity's eternal grace.
    In the realm of relativity, time bends and warps,
Speeding through galaxies, on cosmic ramps.
Black holes devour, with a voracious might,
Bending the fabric, of space and light.
    From Newton's apple to Einstein's mind,
Physics has shaped, the world we find.
From the smallest quark to the grandest star,
Physics unveils the wonder, near and far.
    So let us celebrate, this science divine,
With awe and wonder, in our hearts entwined.
For Physics, dear friend, is the key to explore,
The universe vast, forevermore.

Fifteen

wonders yet to unfold

There lies a science, profoundly divine,
A subject that unravels the mysteries of time.
Physics, the language of the universe's song,
Where equations weave a tapestry strong,
From Newton's laws to Einstein's relativity,
A voyage through space and its vastidity.
Oh, Physics! The art of understanding motion,
From the tiniest quark to the grandest ocean,
With waves that ripple and light that bends,
Unveiling realities that the universe sends.
The laws of thermodynamics, so precise,
Revealing the energy that fuels our lives,
From heat to work, from entropy's sway,
Physics illuminates the forces at play.
Quantum realms, where uncertainty reigns,
Particles entangled, breaking reality's chains,

In superposition, they exist and persist,
Physics dances with mysteries, gently kissed.

Astrophysics, the study of cosmic ballet,
Exploring galaxies in an ever-expanding array,
From black holes devouring with voracious might,
To supernovas exploding, a celestial sight.

Electromagnetism, a symphony of charge,
Binding atoms together, the world at large,
From electricity's spark to magnets' allure,
Physics uncovers the forces that endure.

And so, dear Physics, we bow at your feet,
In awe of the knowledge you help us meet,
A subject of wonder, discovery, and awe,
Unveiling the universe's majestic grandeur.

May we forever delve into your depths,
With curiosity as our guide, we'll take the steps,
To unravel the mysteries you hold,
And witness the wonders yet to unfold.

Sixteen

minds to be blown

Equations dance,
And particles take their chance,
Lies a subject so profound,
Where mysteries are often found.

Physics, the language of the stars,
Unveiling secrets from afar,
From the tiniest atom to the vast expanse,
It explores the universe's intricate dance.

Newton's laws guide our way,
As we study motion day by day,
With every force and every reaction,
Physics sets the stage for cosmic action.

From relativity's bending of light,
To quantum's mysteries shining bright,
Physics delves into the unknown,
Exploring realms we've yet to own.

Electromagnetism's electric embrace,
Binding particles in a cosmic chase,
With Maxwell's equations, we comprehend,
How fields and waves endlessly blend.

Thermodynamics, heat's grand domain,
Reveals the secrets of energy's reign,
From engines to stars, it unravels the might,
Of entropy's dance, day and night.

The laws of optics, in light's embrace,
Paint a picture of the world's sweet grace,
With lenses and prisms, we see anew,
The wonders of colors, both old and true.

Quantum mechanics, the bizarre and sublime,
Challenges reason, yet stands the test of time,
In superposition and entangled states,
Particles defy our normal fates.

Cosmology, the study of space and time,
Unfolds the narrative of our cosmic climb,
From the Big Bang's explosive birth,
To galaxies that span the vastest girth.

Physics, the poetry of the universe,
Unraveling nature's complex verse,
With every equation and every law,
It reveals the beauty that lies in awe.

So let us celebrate this wondrous field,
Where knowledge and wonder are both revealed,
Physics, the gateway to the unknown,
Forever guiding our minds to be blown.

Seventeen

the melody that shapes

Unseen, where wonders hide,
A realm of forces side by side.
Where atoms dance in cosmic space,
Physics unveils its mystic grace.

From Newton's laws to Einstein's dream,
A symphony of motion, it may seem.
Particles collide, electrons spin,
The universe, a stage to begin.

Oh, Physics, you unravel the truth,
Of matter, energy, time, and youth.
Gravity's pull, a celestial might,
Bending stars in the dead of night.

Quantum whispers, uncertainty's call,
Uncloaking secrets, both large and small.
Infinite possibilities, the quantum maze,
Where particles flicker in enigmatic ways.

Black holes devour, space-time bends,
A cosmic dance that never ends.
From galaxies to atoms, the grand design,
Physics, the language that makes them align.

From relativity to quantum's plight,
Physics weaves its fabric, day and night.
In laboratories, minds take flight,
Exploring the cosmos, unlocking its might.

From Isaac Newton to Stephen Hawking,
The torch of knowledge, forever glowing.
Physics, the canvas where genius thrives,
Bridging the gap between truth and our lives.

So let us celebrate the laws that bind,
The universe, with its secrets entwined.
Physics, the melody that shapes our world,
A symphony of knowledge, forever unfurled.

Eighteen

Physics stands tall

Physics stands tall,
A science that unveils mysteries, one and all.
From the tiniest particles to the vast expanse of space,
Physics delves deep, leaving no stone untraced.

Energy and matter, they dance in harmony,
In the laws of nature, they find their symphony.
With equations and formulas, we seek to comprehend,
The workings of the universe, from beginning to end.

Newton's laws guide the motion of every object,
Gravity's pull, we can't help but respect.
Einstein's theory of relativity, a paradigm shift,
Time and space bending, our minds adrift.

From the theory of light to electromagnetic waves,
Quantum mechanics, where uncertainty behaves.
Particles and waves, they blur the line,
In the quantum realm, reality we redefine.

From the atom's nucleus to the stars in the sky,
Physics connects all, with no reason to deny.
From the laws of thermodynamics to the theory of strings,
Physics paints a picture of how the universe sings.

In laboratories, we experiment and explore,
Pushing the boundaries, seeking to know more.
From the Large Hadron Collider to the Hubble's gaze,
Physics unveils secrets in mysterious ways.

So let us celebrate the wonder of Physics,
A subject that sparks curiosity and mystics.
For in understanding the world, both near and far,
Physics shines as a guiding star.

In the realm of science, Physics takes its place,
A discipline that leaves us in awe and amaze.
With every new discovery, we stand in awe,
Physics, the language of nature, forever we adore.

Nineteen

scientific insight

There lies a science like no other,
Physics, the cosmic composer.
From the tiniest particles unseen,
To the vastness of the space serene,
Physics unveils the secrets untold,
Exploring the universe, bold and bold.
Newton's laws, a foundation strong,
Gravity's pull, keeping us along,
Einstein's relativity, bending time's path,
Black holes, where even light meets its wrath.
Electromagnetism, a dance of charge,
Electricity, magnetism at large,
From lightning bolts to the northern lights,
A symphony of energy that ignites.
Quantum mechanics, a mind-bending quest,
Particles entangled, behaving their best,

Uncertainty and superposition, strange and bizarre,
The quantum world, a cosmic memoir.

Thermodynamics, a study of heat,
Entropy's rise, systems retreat,
From steam engines to the laws of heat flow,
Physics reveals how energy will go.

Astrophysics, exploring the cosmic realm,
Stellar birth, death, and overwhelm,
Galaxies swirling in a cosmic ballet,
Physics, the guide, lighting the way.

From the smallest scale to the cosmic expanse,
Physics unravels the universe's dance,
A symphony of equations, theories, and laws,
Revealing nature's secrets, its hidden applause.

So let us salute the wonders of Physics,
A subject that sparks our minds to elicit,
Curiosity, awe, and endless delight,
In the grand tapestry of scientific insight.

Twenty

cosmic hum

In the realm where atoms dance,
Lies the subject of pure chance.
Physics, the mistress of the unseen,
Unveils the secrets of the machine.

From galaxies vast to particles small,
Physics unravels the cosmic sprawl.
Einstein's relativity bends our minds,
As we journey through space and time.

Newton's laws guide the falling rain,
And with every motion, they remain.
Forces, energy, and matter collide,
In equations, we seek truth to confide.

Quantum realms, a mystery untold,
Where particles act both wave and bold.
In quantum entanglement, we find
Strange phenomena that boggle the mind.

From black holes to the cosmic hum,
Physics reveals its symphony, strum.
Through experiments and theories, we roam,
Unlocking the secrets of the unknown.

So let us celebrate this noble quest,
Where knowledge and curiosity are blessed.
Physics, the language of the universe grand,
In its embrace, we forever stand.

Twenty-One

perfect unity

Physics holds its sway,
A realm of wonders, where laws do play.
From tiny particles to stars so bright,
It unveils the mysteries of day and night.
The dance of atoms, their electric charge,
Magnetism's pull, a force at large.
Newton's laws, in motion they reveal,
The secrets of objects, in flight or at rest, they conceal.
Einstein's theory, a grand revolution,
Relativity's grasp, a cosmic solution.
Time and space, intertwined in a dance,
Unraveling the universe, with every glance.
From quantum leaps to the uncertainty,
Particles entangled, in perfect unity.
String theories, dimensions unseen,
A web of possibilities, forever keen.

From the tiniest quarks to the vast expanse,
Physics explores the world's intricate dance.
The laws and equations, a symphony profound,
In the realm of physics, wonders abound.

Matter and energy, entwined and entwined,
A cosmic ballet, in the fabric of time.
From the birth of stars to their final breath,
Physics whispers secrets, defying death.

So let us celebrate this noble pursuit,
The study of nature, in its absolute.
For in the realm of physics, we shall find,
A universe of beauty, forever enshrined.

Twenty-Two

world's contrast

From quantum wonders to laws so grand,
A symphony of knowledge, we understand.
Newton's laws, a foundation strong,
Gravity's pull, a force lifelong.
Einstein's relativity, bending space,
Unveiling secrets, time's elusive chase.
Particles collide in a collider's might,
Revealing the mysteries hidden from sight.
Black holes devour, with gravity's might,
A cosmic dance, in the fabric of night.
From quarks to leptons, a cosmic ballet,
Subatomic realms, where forces play.
Strings vibrate, in dimensions unseen,
The universe's secrets, in physics we glean.
From the tiniest particles to the cosmos vast,
Physics unravels the world's contrast.

Dark matter whispers, in the void it hides,
Through equations and theories, knowledge abides.
From the birth of stars to the death of light,
Physics guides our quest, day and night.
In labs we tinker, with nature we play,
Unveiling the truth, one discovery away.

So let us celebrate, this realm of thought,
Where curiosity and wonder are sought.
Physics, the language of the universe's song,
Forever inspiring, forever strong.

Twenty-Three

❧

transformation and more

Atoms and cosmic might,
Where galaxies dance in eternal flight,
There exists a language, both subtle and grand,
A symphony of numbers, at Physics' command.

From the tiniest quarks to the vastness of space,
Physics unravels the mysteries we chase,
With equations and laws, it seeks to explain,
The nature of reality, in a mesmerizing refrain.

Oh, Physics, you're the maestro of the skies,
Unveiling the secrets that make us wise,
With Newton's laws and Einstein's relativity,
You unravel the tapestry of our connectivity.

In quantum realms, uncertainty prevails,
Particles entangled, where logic often fails,
Yet, in chaos, there's order, a symmetrical art,
Where waves become matter, a mystical start.

From the gentle hum of an electron's song,
To the roar of a supernova, fierce and strong,
Physics, you reveal the universe's design,
A canvas of wonder, where truths intertwine.

The laws of thermodynamics, unyielding and sound,
Teach us that energy cannot be lost, but found,
From heat to motion, transformation and more,
Physics guides us through the fundamental lore.

Through telescopes and colliders, we explore,
Venturing into the unknown, forevermore,
Seeking answers to questions that lie untold,
In the depths of space, where mysteries unfold.

Oh, Physics, you're the bridge to our dreams,
Connecting the realms, or so it seems,
With quantum leaps and cosmic rays,
You illuminate the path where knowledge plays.

So let us celebrate, with minds astir,
The wonders of Physics, forever to stir,
For in understanding the laws of this land,
We grasp the essence of our existence, hand in hand.

Twenty-Four

❦

quest for truth
and reason

Cosmic skies,
Where particles dance and theories arise,
There lies a realm of knowledge profound,
Where Physics reigns, with wisdom unbound.

From Newton's laws to Einstein's relativity,
Physics unveils nature's exquisite creativity.
Quantum leaps and waves that bend,
Unveiling secrets, transcending the end.

Oh, Physics, the language of the universe.
With equations and formulas, you immerse,
In the depths of space and the fabric of time,
Unveiling mysteries, both complex and sublime.

From the tiniest quarks to galaxies afar,
Physics explores the cosmos, a guiding star.

Electricity, magnetism, forces at play,
Unraveling the laws that govern each day.
From the fusion of stars to the birth of light,
Physics reveals the secrets, shining so bright.
Black holes devouring, with infinite might,
Expanding horizons, illuminating the night.
From the wonders of sound to the colors we see,
Physics explains the world, a symphony.
Harmonizing nature, in perfect balance,
Unveiling the mysteries, with elegant grace.
Oh, Physics, the quest for truth and reason,
A constant pursuit, through every season.
From laboratories to the vast unknown,
Physics, the foundation on which knowledge is sown.
So let us celebrate this subject of might,
With curiosity, let our minds take flight.
For in Physics, we find beauty and awe,
A universe of wonders, forever in awe.

Twenty-Five

a realm of uncertainty

Numbers and equations, I find delight,
Where the mysteries of the universe come to light.
Physics, a symphony of forces and energy,
Unveiling the secrets of our cosmic tapestry.

From the tiniest particles, so infinitesimally small,
To the vast expanse of space, where galaxies sprawl,
Physics unravels the fabric of reality's design,
With theories and laws that make the universe shine.

Newton's laws, a foundation on which we stand,
Gravity's embrace, guiding objects across the land.
Einstein's relativity, bending time and space,
Revealing the wondrous dance of matter in its embrace.

Quantum mechanics, a realm of uncertainty,
Where particles flicker in a quantum duality.
In the quantum world, where waves and particles entwine,
Physics challenges our perceptions, expanding our mind.

From the delicate balance of atoms in motion,
To the explosive power of nuclear fusion,
Physics weaves the threads of matter and energy,
Unveiling the symphony of the cosmic symphony.

From the gentle flutter of a butterfly's wings,
To the swirling storms that chaos often brings,
Physics seeks to understand the patterns we see,
In the intricate dance of nature's tapestry.

So let us celebrate the wonders of this science,
Where curiosity and exploration intertwine,
For in the realm of Physics, we find our place,
Amidst the grandeur of the cosmic embrace.

Twenty-Six

for all humankind

Science, where wonders unfold,
There lies a subject, mighty and bold.
Physics, the language of the universe divine.
Unraveling mysteries, layer by layer, line by line.

From tiny atoms to galaxies afar,
Physics reveals the secrets they bizarre.
Newton's laws, with elegance and grace,
Guiding motion in every time and space.

Electromagnetism, a force so grand,
Binding particles with an invisible hand.
Electric charges dance and sway,
Creating waves of light, in a cosmic ballet.

Oh, the beauty of quantum mechanics,
Where particles behave in ways so dynamic.
Uncertainty reigns, in a quantum haze,
Yet, through Schrödinger's equation, we gaze.

Relativity, the theory profound,
Bending time and space, a spectacle renowned.
Black holes devour, with a gravitational pull,
The fabric of reality, a cosmic whirlpool.

From the tiniest quarks to the vast expanse,
Physics weaves a tapestry, with elegance.
Exploring the depths of nature's might,
Unveiling the laws that govern day and night.

So let us celebrate this wondrous art,
With curiosity and an open heart.
For in the realm of physics, we find,
A journey of discovery, for all humankind.

Twenty-Seven

fabric of our world

Reason thrives,
And knowledge takes its noble strides,
Lies a subject that forever strives,
To unravel the universe's hidden sides.

Physics, the art of understanding,
The laws that govern every landing,
From the tiniest atom to the vast expanse,
It delves into the cosmic dance.

With equations that dance on the page,
Formulas that unlock nature's cage,
The world of matter, energy, and light,
Becomes a canvas for scientific insight.

From Newton's apple, a moment of grace,
To Einstein's relativity, a paradigm's embrace,
The wonders of physics never cease,
Unveiling secrets, granting us peace.

From the speed of light that knows no bounds,
To the quantum realm, where uncertainty abounds,
Particles that entangle and entwine,
Creating a tapestry both complex and divine.

In laboratories, minds ignite,
Exploring the mysteries, day and night,
From particle colliders that smash and collide,
To telescopes that capture the stars' pride.

Magnets attracting with invisible might,
Creating forces that bind day and night,
Gravity's pull, keeping worlds in sync,
The laws of physics, the universe's link.

In the elegant language of mathematics,
Physics speaks, revealing its tactics,
From motion and forces to waves in the sea,
Its symphony resonates, setting knowledge free.

So let us celebrate this wondrous field,
Where imagination and reason yield,
Let us marvel at the wonders it has unfurled,
The beauty of physics, the fabric of our world.

Twenty-Eight

the mysteries intertwine

In the realm of atoms and galaxies,
Where mysteries unfold through cosmic seas,
Lies a science that unveils nature's laws,
A symphony of equations, a physicist's cause.
Physics, the language of the universe,
Through theories and experiments, it traverses,
From the smallest quark to the vast expanse,
It unravels the secrets of every circumstance.
With Newton's laws, we understand motion,
The force that guides every object's notion,
From falling apples to planets in flight,
Physics brings clarity to nature's might.
Einstein's relativity takes us far and wide,
Bending time and space, a cosmic ride,
The fabric of reality, twisted and turned,
In a dance of matter, energy, and what we've learned.

Quantum mechanics, a realm so strange,
Particles entangled, their behavior deranged,
A wave-particle duality, a quantum embrace,
Uncertainty reigns, in this ethereal space.

From thermodynamics to electromagnetism,
Physics shapes our world with profound realism,
From light's speed to sound's vibrations,
We marvel at the laws of nature's creations.

In labs and observatories, scientists delve,
Pushing boundaries, seeking knowledge to excel,
Physics, a quest for truth, an endless pursuit,
A tribute to curiosity and the wonders it can compute.

So let us celebrate this noble science,
With awe and reverence, let us commence,
For physics reveals the grand design,
And in its embrace, the mysteries intertwine.

Twenty-Nine

⤜⦿⤏

journey of discovery

There lies a realm of mystery and awe,
A subject that leaves us in awe-struck awe.
Physics, the art of understanding the universe,
Unveiling the secrets that lie in every verse,
From the tiniest particles to galaxies far,
Physics unravels the wonders that are.
From Newton's laws that govern our world,
To Einstein's relativity, beautifully unfurled,
Quantum mechanics, a realm so bizarre,
Where particles teleport and mysteries mar.
Oh, Physics, you guide us through the unknown,
With every equation and principle shown,
From waves and fields to energy and light,
You illuminate our minds with pure delight.
In laboratories, we experiment and strive,
Uncovering truths that make us feel alive,

From the laws of motion to electromagnetism,
Physics shapes our lives with profound rhythm.
From the stars that twinkle in the night,
To the waves that crash in morning light,
Physics surrounds us in every way,
A symphony of knowledge that will never sway.
So let us celebrate this noble art,
That weaves together logic and the heart,
For in the study of matter and its motion,
Physics reveals the beauty of our existence's notion.
Through equations and theories, we explore,
The universe's secrets that lie at its core,
Oh, Physics, you are a gift so divine,
A journey of discovery that forever will shine.

Thirty

unwavering pride

Stars afar,
Lies the beauty of Physics, a guiding star.
From the tiniest particles, so mysterious and small,
To the grandest cosmos, captivating us all.

Newton's laws, they teach us about motion,
For every action, there's an equal reaction.
Gravity, a force, pulling all things down,
Binding the universe, with an invisible crown.

Einstein's theories, they bend time and space,
Relativity's wonders, a mind-bending chase.
The speed of light, a cosmic speed limit,
Exploring the universe, we're all committed.

Quantum mechanics, a world so strange,
Particles dancing, in a quantum exchange.
Uncertainty, entanglement, and wave-particle duality,
Challenging our minds, with endless curiosity.

Thermodynamics, the study of heat,
Flowing from hot to cold, so discrete.
Entropy, a measure of disorder's reign,
The second law, it cannot be restrained.

Astrophysics, exploring the celestial sphere,
Stellar evolution, galaxies so clear.
Black holes, pulsars, and cosmic rays,
Revealing the secrets of our cosmic maze.

From electromagnetism to nuclear might,
Physics unveils nature's hidden light.
From the laws that govern the tiniest domain,
To the vastness of space, where galaxies reign.

Physics, oh Physics, a subject so grand,
Expanding our knowledge, like shifting sand.
With every discovery, a new door opens wide,
Unveiling the universe, with unwavering pride.

Thirty-One

scientific art

In the realm of atoms, where mysteries lie,
Physics unravels the secrets of the sky.
From subatomic particles to galaxies afar,
It unveils the wonders, both near and far.

Oh, Physics, the language of the universe,
With equations and laws, you immerse.
From Newton's apple to Einstein's relativity,
You guide us through this cosmic activity.

In the realm of forces, where dynamics unfold,
Physics explains how objects behave and hold.
Gravity pulls, magnetism attracts,
Electromagnetic waves in space interact.

Quantum mechanics, a puzzling delight,
Particles entangled, dancing in the night.
Uncertainty reigns in this quantum domain,
Yet Physics persists, unraveling the chain.

From the laws of thermodynamics to energy's flow,
Physics illuminates the world we know.
From the pendulum's swing to the light's refraction,
It brings clarity to nature's interaction.

Cosmology, the study of the vast unknown,
Physics explores what lies beyond our own.
Black holes and supernovae, expanding space,
Revealing the secrets of time and place.

Oh, Physics, the cornerstone of knowledge profound,
Unveiling the mysteries that surround.
With every discovery, a new chapter unfolds,
Expanding our minds as the universe unfolds.

So let us celebrate this scientific art,
Embrace the wonders that Physics imparts.
For in understanding the laws that govern us all,
We find beauty in this cosmic ball.

Thirty-Two

Physics reigns

Physics, a symphony of laws,
Unraveling the universe's cause.
From the tiniest particles unseen,
To galaxies vast, majestic and serene.
Physics unveils the cosmos' dance,
Guiding us through its cosmic expanse.
Newton's apple, falling from the tree,
Einstein's relativity setting minds free.
Quantum mysteries, particles entwined,
String theories weaving what we find.
Energy, matter, motion, and light,
Mysteries that keep us up at night.
From the laws of motion, we derive,
The beauty of nature, we strive to describe.
From the gentle breeze to the stormy gale,
Physics reveals the forces that prevail.

Electromagnetism, a dance of charge,
Atoms colliding, rearranging at large.
    Black holes lurking, bending space,
Stars exploding, leaving no trace.
Gravitational waves ripple through time,
Unveiling secrets, so sublime.
    From the depths of the atom's core,
Nuclear forces, forever more.
Fusion, fission, a power untamed,
Harnessing energy, we have aimed.
    The cosmos whispers its cosmic tale,
Physics listens, never to fail.
From galaxies colliding to the birth of stars,
Physics reveals the grandest memoirs.
    So let us celebrate this science so grand,
With awe and wonder, hand in hand.
For in the realm where knowledge unfolds,
Physics reigns, a story yet untold.

Thirty-Three

intricate bend

Lies a kingdom of knowledge, a realm of delight,
The subject that unveils the universe's might.

Physics, the oracle of the natural domain,
Unraveling mysteries with each tiny grain,
From the tiniest quark to the grandest star,
It reveals the secrets that make us who we are.

Through formulas and theories, it paints the scene,
Exploring the forces that shape and careen,
Gravity's embrace, bending light's flight,
Electromagnetic waves, glowing bright.

From Newton's apple to Einstein's relativity,
Physics weaves tales of celestial activity,
Black holes devouring, time's intricate bend,
Quantum leaps, where the rules transcend.

In labs and observatories, scientists delve,
Unraveling the fabric, they refuse to repel,

Colliders collide, particles collide,
Unveiling the universe, side by side.
 From the laws of motion to thermodynamics' heat,
Physics explains how the world does repeat,
From Newton's cradle to a pendulum's swing,
The laws of physics, forever reigning.
 The symphony of atoms, the dance of the spheres,
Physics reveals nature's harmonious years,
From the birth of the cosmos to its final breath,
Physics guides us through life and death.
 So let us celebrate this noble pursuit,
Where curiosity and wonder take root,
For in the realm of equations, a truth unfolds,
That physics, the language of nature, never grows old.

Thirty-Four

Physics reigns supreme

In the realm where logic meets the stars,
Lies a realm that captivates and jars,
Physics, the mistress of the unknown,
Unveiling secrets, layer by stone.

From particles unseen, to cosmic dance,
Exploring the universe, at every chance,
Newton's laws and Einstein's theories,
Unraveling nature's deepest mysteries.

Electromagnetism, a force so strong,
Binding atoms, singing nature's song,
Quantum leaps and waves that collide,
In black holes where time and space coincide.

From the smallest atoms to the galaxies vast,
Physics reigns supreme, unsurpassed,
It shapes our world, both near and far,
Guiding us through the cosmic memoir.

From the depths of space to the subatomic fray,
Physics sheds light on the night and day,
From the laws of motion to the theory of strings,
It's a symphony of knowledge that physics brings.

So let us celebrate this noble quest,
To understand the universe's behest,
For in the realm where logic meets the stars,
Physics reigns supreme, dispelling all memoirs.

Thirty-Five

speed of light to gravitational waves

Of laws and equations,
Where matter dances and energy awakens,
There lies a quest for truth profound,
In the realm of physics, where wonders abound.
From Newton's apple falling from the tree,
To Einstein's theory of relativity,
Physics unravels the secrets of space,
Unveiling the mysteries of time and place.
In the quantum world, where particles collide,
Uncertainty reigns, and waves subside,
Particles entangled in a cosmic ballet,
Quantum mechanics leading the way.
From atoms to galaxies, the universe is vast,
And physics explores it with a steadfast grasp,

From the tiniest particles to the grandest of scales,
Physics uncovers the secrets that nature unveils.
From the laws of motion to the laws of thermodynamics,
Physics delves into the world of mechanics,
It measures the forces that shape our world,
And explains the phenomena that unfurl.
From electricity's spark to magnetism's pull,
Physics illuminates the world in full,
From the speed of light to gravitational waves,
Physics guides us through the cosmic maze.
With every discovery, a new door unlocks,
Revealing the wonders that nature concocts,
Physics, the language of the universe we seek,
Unraveling the mysteries, week after week.
So let us celebrate this noble science,
With curiosity and a sense of defiance,
For in the world of physics, we find our place,
Exploring the cosmos with wonder and grace.

Thirty-Six

curiosity intertwine

Forces play,
There lies a science that holds us in sway.
Physics, the art of understanding the laws,
Unveiling nature's secrets and her hidden flaws.

From the tiniest particles that make up all,
To the grandest cosmos that leave us in awe,
Physics unravels the mysteries of time,
With equations and theories, sublime.

Newton's laws guide us through motion's embrace,
As objects in harmony find their rightful place.
The apple that fell, a revelation profound,
Gravity's force, the Earth's pull, now renowned.

Einstein brought forth a revolution so grand,
With relativity, time and space expand.
The bending of light, a celestial ballet,
Black holes and wormholes, a cosmic display.

Quantum mechanics, a realm so bizarre,
Where particles vanish and entangle from afar.
Uncertainty reigns in this quantum domain,
With wave-particle duality, a puzzle we can't explain.

From thermodynamics to electricity's spark,
From optics to magnetism, a symphony so stark.
Physics weaves a tapestry of laws and theories,
Connecting the world, unlocking its mysteries.

In labs, we explore, experiment, and create,
Pushing the boundaries, defying fate.
From atoms to stars, the universe we explore,
Physics, the language that leaves us in awe.

So let us celebrate this wondrous domain,
Where knowledge and curiosity intertwine.
Physics, the art that enlightens our way,
Forever inspiring, from night to day.

Thirty-Seven

worth more than gold

When wonders unfold,
There is a science that's timeless, worth more than gold.
Physics, the language that nature obeys,
Unveiling the secrets of cosmic displays.

From the smallest of particles to galaxies afar,
Physics unravels the mysteries that are.
Quantum mechanics dances with uncertainty,
While relativity brings time to a knee.

With equations and formulas, we explore the unknown,
From Newton's laws to Schrödinger's own.
Einstein's theories bend our perception,
While Planck's constant fuels quantum conception.

In labs filled with marvels and experiments grand,
Scientists delve into the laws of the land.
From the motion of planets to the flight of a kite,
Physics reveals the beauty of day and night.

From optics to thermodynamics, a captivating blend,
The laws of physics have no end.
Electromagnetism sparks the world's connection,
While nuclear forces bind with perfection.

The universe sings its symphony of light,
As physics unveils the secrets of its might.
From the birth of stars to the black holes' embrace,
Physics paints a picture of infinite space.

So let us celebrate this wondrous quest,
Where science and curiosity are truly blessed.
Physics, the language that speaks to our soul,
Unveiling the universe, making us whole.

Thirty-Eight

mistress of the universe

Physics, the mistress of the universe's laws,
Unveils the secrets, with relentless applause.
From the tiniest particles, so incredibly small,
To the vastness of space, where galaxies sprawl,
Physics unravels the mysteries of time and space,
Guiding us through the cosmos, at an exhilarating pace.
Oh, the elegance of equations, so precise and neat,
Calculating the forces that make our world complete.
Newton's laws, with their profound clarity,
Describe the motion of all things with rarity.
Einstein's theory, with its relativity profound,
Unveils the secrets of gravity, tightly wound.
Bending the fabric of space, like a cosmic trampoline,
Revealing a universe, like we've never seen.
Quantum mechanics, a realm of uncertainty,
Where particles teleport, with eerie frequency.

Wave-particle duality, a paradox so grand,
Challenging our notions of how the universe is planned.

From thermodynamics to electromagnetism's might,
Physics enlightens us, like a beacon of light.
Oh, the wonders of electricity, with currents that flow,
Powering our world, with a mighty glow.

From the intricacies of light, with its vibrant spectrum,
To the wonders of sound, vibrating through the diaphragm,
Physics captures the essence of our sensory delight,
Transforming energy, with all its might.

So let us celebrate this science, with hearts aglow,
For Physics, the mistress, has so much to show.
In the pursuit of knowledge, we shall forever roam,
For Physics, the universe's language, is our eternal home.

About the Author

Walter the Educator is one of the pseudonyms for Walter Anderson. Formally educated in Chemistry, Business, and Education, he is an educator, an author, a diverse entrepreneur, and the son of a disabled war veteran. "Walter the Educator" shares his time between educating and creating. He holds interests and owns several creative projects that entertain, enlighten, enhance, and educate, hoping to inspire and motivate you.

Follow, find new works, and stay up to date with
Walter the Educator™
at www.WaltertheEducator.com

Milton Keynes UK
Ingram Content Group UK Ltd.
UKHW020830050923
428087UK00016B/1052